Barbie®

Four Decades In Fashion

Introduction by **Barbie**

as told to Laura Jacobs

A MINISERIES BOOK

Abbeville Publishing Group • New York • London • Paris

Front cover: 1998. Corduroy overalls. See page 187.
Back cover: 1977. Ice-blue satin gown with jewel-encrusted bodice and skirt; matching satin shawl with pale pink lining; crystal drop earrings.
Spine: 1977. Jacket over slim-cut trousers. See page 181.
Frontispiece: 1996. Sleeveless white silk gown. See page 175.

Editor, 1st ed.: Renée Klock
Editor, 2d ed.: Jeffrey Golick
Designer: Molly Shields
Production Manager: Lou Bilka

Barbie and associated trademarks are owned by and used under the license from Mattel, Inc. All Barbie doll images and art are owned by Mattel, Inc. Copyright © 1999 Mattel, Inc. All rights reserved.

Copyright © 1999 Abbeville Press. Compilation, including selection of text and images, copyright © 1999 Abbeville Press. All rights reserved under international copyright conventions. No part of this book may be reproduced or utilized in any form or by any means, electronic or mechanical, including photocopying, recording, or by any information storage and retrieval system, without permission in writing from the publisher. Inquiries should be addressed to Abbeville Publishing Group, 22 Cortlandt Street, New York, N.Y. 10007. The text of this book was set in BeLucian Book and Metropolis. Printed and bound in Hong Kong.

First edition
10 9 8 7 6 5 4 3 2 1

The Library of Congress has cataloged the Tiny Folio edition as follows:
Barbie: four decades of fashion / introduction by Barbie as told to Laura Jacobs. — [Rev.] 2nd ed.
   p. cm.
   "A tiny folio."
   ISBN 0-7892-0461-4
   1. Barbie dolls—Clothing. I. Title.
NK4894.3.B37B38 1998
Rev. ed. of: Barbie : in fashion. 2nd ed. 1994.
688.7'221—dc21   98-8808

MiniSeries ISBN 0-7892-0552-1

Contents

A Life in Fashion
by **Barbie**, as told to Laura Jacobs * 7

1959 and the **Sixties** * 19

the **Seventies** * 61

the **Eighties** * 103

the **Nineties** * 145

Appendices * 188

A Life in Fashion

by **Barbie**

as told to Laura Jacobs

I've never been really good at talking about myself. I think actions speak louder than words, and ever since I first started gaining notice as America's most famous Teenage Fashion Model I've always tried to put my best foot forward. Leafing through this Tiny Folio, a photo album of all the fabulous fashions I have worn through the years, I suddenly realize how much has happened during my lifetime. It's been a high-flying roller coaster ride, and you, my wonderful friends and fans, have enjoyed the thrills and chills with me—all the time inspiring me, confiding in me, caring for me. I couldn't have done it without you.

    Today, models have become big business. Like actresses and entertainers, fashion models are superstars with incredible incomes. But when I began my modeling career in March 1959, the world was a different place. The stars were up in the sky— and in our eyes. After I answered "yes" to my first photo assignment, I hung up the phone and jumped for joy. When I floated back down to earth, I went straight to my desk to look up the word "model" in the dictionary.

    "A miniature representation of something; an example for imitation; an ideal." That definition in Webster's told me almost more than I wanted to know about the career path I'd begun. Of course it was thrilling to be the first girl in high school to earn a real live paycheck—simply for wearing the most beautiful clothes in the world! But now all eyes would be watching *me*. Girls I didn't even know would look up to me. I decided I wasn't going to let them down.

Right from the start, I was known by just my first name. In fashion, this is a time-honored tradition. If a model is lucky, her name will be a perfect match for her look and character, almost symbolic. For example, in my first decade of modeling, the 1960s, there was huge Veruschka, an endlessly exotic Russian, and skinny Twiggy, a tender British shoot. As you can see, compared to these, Barbie is a rather normal name. But it turns out that I was in the right place at the right time, because fashion was ready for a girl who could represent all-American can-do and enthusiasm.

There's no question that my full figure caused a sensation. Some people say I was America's answer to France's yé-yé girl, Brigitte Bardot. But I've always been more apple pie than cheesecake. In 1959, the year I debuted, the hourglass shape was the ideal, and the many minutes a girl spent on her appearance were considered an important part of self-image. Eyeliner, lipstick, and a powdered nose, undergarments that smoothed and shaped the body into predictable curves—these were the foundations of an ensemble. An ensemble style was the key to elegance. And elegance was the result of discipline. In the early 1960s, all a teen had to do was look to America's

First Lady, Jacqueline Kennedy, to see that only out of effort came effortlessness. My look was curvier than the First Lady's, yet we both wore ensembles of impeccable design, proving that a beautifully draped and tailored garment looks good on more than one body type.

Beauty Treatments

When I first began modeling you could say that makeup was a bit more *emphatic*. Eyebrows were carefully sculpted and for two years mine arched in a cool, sophisticated manner. Eyeliner, especially, required a super calm hand (I took a deep breath before drawing each line), and for many years I preferred an almond-eyed tilt, very Sophia Loren. Suntanning was definitely NOT IN, and my skin was an ivory bisque.

Early in the sixties, though, my makeup artist relaxed my brows and toned down my lip and nail shades to a pale pink (a sort of strawberry ice cream), and we began highlighting the rosy tones in my complexion. In 1971, during a trip to Malibu, I got my first tan. If you look through my portfolio — in fact, not just mine but that of any model of my generation — you will notice a softening of rigid rules in makeup.

Oh, sure, there will always be flights of fancy — in fashion, that's part of the fun — glitter and pastels during the "mod" years, and some wild eye shadows when I toured with the Rockers in the 1980s. But when I look through my book (that's fashion lingo for portfolio) I see a transformation from a very exact and formal self-presentation, an almost theatrical stress on perfection (and you must remember, in those days my boyfriend Ken and I were very involved in our Little Theatre group), to a more informal, natural kind of beauty. When I look at my early self, I see a young woman who seems to be keeping secrets. The way I look today—smiling so much you can see my dimple!—strikes me as more direct and ready for anything.

Even my hairstyles reflect the trend toward brush-and-go preparation. Once I let my hair down in the late sixties, I never really went back to fancy coiffeurs. From the seventies on, I've worn my locks long and wavy, and have settled on blonde as the color that suits me best (though I still love to experiment). But for many years I had a handful of classic colors—White Ginger, a creamy platinum; and Titian, a Renaissance auburn—and distinctive hairstyles that came to be identified with me. Say the name "Barbie" and people still think of my very first "do": that famous swing ponytail with the poodle bangs. I practically pranced when I wore it. There's also my Bubble Cut (so smart!), Page Boy (so sleek!), and the big flip I tried in 1965 called American Girl (move over Ann-Margret!). Probably the most glamorous style from my early years was the Swirl Ponytail, gleaming bangs swept boyishly across my brow. Pure Park Avenue meets Hollywood, it made me sashay like a movie star.

Some people cringe when they catch glimpses of "the way they were," but I adore how a hairstyle or cosmetic color brings back the atmosphere of an era. Chocolate Bon Bon and Cupa-Co-Co are not just guilty pleasures that test my waistline, they're two of my haircolors from 1969, and they conjure up the plain-spoken sensuality that started the seventies (remember, *Hair* didn't hit Broadway until 1968). An even better form of remembrance is the touch and sight of the clothes one wore. The feel of the fabric. The character of the construction. The line, the length, the LOOK. Boy, did I wear some beauties. Say "Barbie," and people will describe fashion favorites from as far back as '59—as if it were yesterday!

High Fashion Fifties

As I said earlier, my timing couldn't have been better. When I took my first, dare I say nervous, turn down the runway, fashion was still measured in the magical, manicured hands of this century's great international houses: Balenciaga, Chanel, Dior. It was a time of couture classicism and my early ensembles spoke with a daring French accent, and sometimes a little Italian. My first modeled item was the simplest, but what a whirl of associations it had: that jazzy black-and-white striped swimsuit, worn with sunglasses and gold hoop earrings, was chic beyond words—Coco Chanel in Cap Ferat, Gina Lollobrigida on the Riviera. (Comforting too—made of knit jersey that felt fuzzy and cozy against my skin.) It may now be the most instantly recognizable bathing suit in history.

Other unforgettable ensembles from that first year? I must mention my blue-and-white polka-dot pouf dress with snow-bunny stole, at once extravagant and pristine, as if wondering, "Which way to Maxim's?"; a navy blue knit two-piece suit, the ultimate office elegance, yet with a red petal hat that carried one regally into evening; and the first of my many superbly cut sheath dresses (a style that seemed to add inches to my height), cloaked in black faille with big, bowed patch pockets. Wearing that coat was like being wrapped in a timeless, moonless night.

And yet change was in the air. Ballet dancer Rudolf Nureyev defected from the Soviet Union in 1961, the Twist untwisted decorum in '62, Valentina Tereshkova became the first woman launched into outer space in '63, and in their collarless suits, the Beatles blasted off on Ed Sullivan in 1964. Everyone, everything was getting aired out. My beautiful ensembles of the early sixties, such a pleasure to wear, yet requiring such planning and unflinchingly perfect posture, gave way to kickier, more youthful designs. The "ensemble" as we knew it (coordinated shoes, gloves, handbag, and hat), loosened its grip on our lives.

Sizzling Sixties and Sensible Seventies

And we began to move differently. British and American designers gave us clothes that danced—the Frug, the Pony, the Mashed Potato—and I wore, or rather, danced them all. The Mod years were here, the miniskirt was now, and with it a more animated, angular, energized way of posing, romping, *being*! Under a rainbow of phosphorescent colors, I loosened up too, found a freer, more athletic movement style, and began to wear lower heels. While the Beatles sang about "Lucy in the Sky with Diamonds," my diamonds were on my new "Color Magic" swimsuit, a harlequin pattern of hot yellows, greens, and blues. In just six years we'd gone from quintessential black and white to the psychedelics of Peter Max. Peek into my mod closet of madcap concoctions and you'll find a mini modern museum of op-art geometrics, neon colors mixed with squares of silver. My midi-length shirtdress from 1969, with its tangerine zigzags on a field of gold lamé, sums up the dizzy electricity of the era.

Am I upsetting anyone if I say that the 1970s weren't my favorite decade for design? The force field fizzled, and the seventies saw a regrouping, a rethinking, a return to grassroots and groundswell. Ecology and economy were the bywords of the time and in fashion terms that meant knits, linens, wools, in league with synthetics of all kinds. Granny and peasant dresses, crocheted vests, lots of denim, and pants, pants, pants were everywhere. My standouts from these years were my evening gowns, stunning creations in chiffon, paisley lamé, and embroidered georgette.

Extravagant Eighties

In 1980, the movie *10* (as in "a perfect 10") came out and to my mind, it set the tone of the decade. Women were ready to do and have it all. To be or not to be perfect wives, mothers, and career professionals—all at once—that was the question. The women characters on popular nighttime soap operas like "Dallas" and "Dynasty," whether they were CEOs or MRSs, glittered in glamorous gowns and power suits, examples of a new female elite. This stress on grasping and sculpting our own destinies went hand in hand with the fitness trend, which set a new standard for female strength and shape. Muscle was in; indolence was out; "definition" was the watchword. It wasn't just upper arms and midriff that we were redefining, but our place in society.

This obsession with fitness did not go unnoticed, or unrewarded, by designers. Stretch Lycra® and wool jersey became favored fabrics, and old-style architectural seaming gave way to a fluid, body-conscious honesty. Our figures shaped the fabric. A Lycra® skirt was now like a fingerprint—it didn't fit two bodies the same way. Aerobic fitness was the new girdle.

Although I have always been both active and careful about my figure (at 5'9", my happiest weight is 110 pounds, and I've been consistent about it), I too caught the fitness fever, and in 1986 I launched my own line of "Barbie 'B' Active Fashions," a series of snappy sweatsuits and sportif-inspired daywear. A capital "B" was my designer logo. In the eighties I also wore a line of clothing that really understood the busy, all-systems-go pace of the American woman; Spectacular Fashions offered a new kind of ensemble: a multipiece color-coordinated wardrobe of five fabulous looks. Why hadn't anyone thought of this before?

Nostalgic Nineties

A curious thing happened to me in the 1990s. I became an "icon." That's a fancy word intellectuals use when they want to say something or someone is both famous and symbolic. Which is what happened to me in 1994, when I reached the thirty-fifth anniversary of my modeling career. Suddenly everyone was writing books and articles about me: remembering my debut in 1959, analyzing my figure and my fashion influence, and honoring my amazing popularity and staying power. Parties were thrown for me all around the world. It was exciting and touching.

Another curious thing happened in the 1990s. I noticed that, more and more, my fans wanted to see me dress up like other icons. I think this showed the influence of the rock star Madonna, who was always pretending she was someone else, toying with the idea of "identity" (that's another word intellectuals like). Sure, I continued modeling contemporary clothes of all kinds, from the high fashion of Bill Blass and Escada to the newest beach clothes and Olympic gym wear (the nineties are nothing if not eclectic!). But I just as often found myself in the famous outfits of historical and Hollywood figures like Madame du Barry, Scarlett O'Hara, Eliza Doolittle, and Dorothy from *The Wizard of Oz* (Ken was the Tin Man).

More than once I was costumed as a mermaid! Just between us: this decade is obsessed with mermaids. Luckily, I've always been a super swimmer.

In the Pink

Anyone who knows me knows that my signature color is pink. I like to think its temperament is akin to mine, and I have worn the color with joy throughout the years. A soft blush was my favorite shade early on, and a lacy, pale pink dotted-swiss dress from my first year is, to me, heartbreakingly pretty. Two years later, I modeled a whip-cream swirl of a gown, probably the most feminine creation I've ever worn, and a masterpiece of design. When I won Teen Journal's essay and photo contest in 1962, and spent the summer in New York working at the magazine as a teen model and Special Guest Editor, I learned that pink is one of the most flattering shades against the complexion (that's why a pink umbrella can be such a great accessory, and some of my best hats have been pink).

Cotton-candy pink entered my wardrobe in the 1960s with a satin gown and matching marabou boa that tickled my nose, and a superb dressmaker suit trimmed with pink satin frosting—Parisian savoir faire! My most beloved outfit from the 1970s was a pink-and-white gingham gown with organdy sleeves and black velvet bows at the neck and wrists. I keep it in tissue today (pink, of course!). And a much loved sweater is my curly cowlneck from 1986, the color of bubblegum. Hot and shocking pinks, and many lavenders (the fairy-tale cousin to pink), entered my wardrobe in the eighties and are still going strong in the 1990s. Pink is like family to me.

All That Glitters

Dazzle has always been a part of my mystique, but as you can see looking through this book, the sparkle has usually been in my step or on my dresses—just feast your eyes on the magnificent Bob Mackie gowns I started wearing in 1991, beaded sequined wonders. Oh, on occasion I have worn diamond earrings, and in the 1980s I went with the flow and wore fun faux jewelry (I'm absolutely mad about the little green palm tree earrings that went with my polka-dot bikini in 1988). My first gold hoops are now classics, and I would hate to lose one. But of all the gems one might choose from, pearls have been my constant companion. I treasure their modesty, their mystery, their luster, their life. If I were a jewel, I like to think I would be a pearl. A pink one.

1959 and the Sixties

1959
Sheath dress with blue skirt and red-and-white-striped top;
matching striped coat with decorative gold buttons

1959
Tailored red-and-white-checked cotton bodysuit and
"clam digger" denim jeans

1959
Salmon sailcloth sundress
with white chef's apron

1959
Pink dotted-swiss dress; white petticoat trimmed
with tulle ruffle and ribbon bow

1959
Blue taffeta bubble dress with white polka dots;
faux rabbit-fur stole lined with white satin

1959
Strapless gold brocade sheath dress; matching coat
lined in satin and trimmed with faux mink

1959
Black-and-white-striped swimsuit

1959
Baby Doll nightgown and panties trimmed
with embroidery and satin bows

1959
Classic blue knit suit with tailored blue-and-white-checked rayon bodyshirt

1959
Apple-print sheath dress; black faille coat
with gathered yoke and patch pockets

1959
Strapless satin wedding gown with long-sleeved tulle overlay and three-tiered skirt

1959
Blue-and-white cotton sundress

1959
Red sailcloth jacket trimmed with white braid;
white cotton short shorts; striped cotton T-shirt

1960
Full-skirted powder-blue corduroy jumper with felt appliqué;
white cotton puffed-sleeve blouse

1960
Strapless glittery black sheath gown with tulle flounce;
pink chiffon scarf

1960
Red linen suit with sheath skirt
and red-and-white-striped blouse

1961
Regal pink satin gown with flowing side-draped train;
faux rabbit-fur stole lined in pink satin

1961
Ballerina tutu of white net over
silver lamé bodice

1961
Red cotton sheath dress with cap sleeves, patch pockets,
and decorative gold buttons

1961
Short-sleeved sweater, striped cotton capri pants,
and poplin car coat with toggle fasteners

1962
Black taffeta sheath dress with cap sleeves
and matching taffeta bow at neckline

1962
Red velvet coat lined in white satin

42

1963
Sheath dress of red and gold lamé with matching coat lined in red
and trimmed with faux fur

1963
Persimmon jumpsuit with gold net hostess coat

1963
Emerald green satin suit with shawl-collared jacket and sheath skirt
with peplum flounce; sleeveless white satin blouse

1963
Satin wedding gown with flowing chiffon overskirt
and ruffled front insert

1964
White satin evening gown with tulle overskirt in alternating panels of pink and red

1964
Strapless black sheath dress with black ribbon trim
and sheer tulle cape

1965
Green organza and floral-print cotton dress with shawl collar
of green organza and matching organza sash

1965
Blue linen jacket with shawl collar edged in emerald green satin;
sleeveless dress with antique silk print top and linen sheath skirt

1966
Bone double-breasted leather-look coat

1966
Pink sheath dress with satin top and cotton hopsack skirt; matching hopsack jacket with satin trim and front tie

1966
Pink satin sheath dress with overdress of sheer white lace
and rose satin waistband with bow

1967
Wedding gown with sweetheart bodice and flared skirt revealing
satin slip with braid trim and dotted tulle overskirt

1967
"Wet Look" red vinyl coat with white stitching,
lined in blue taffeta

1967
A-line dress of yellow, green, and red panels
trimmed with black braid

56

1968
Red velvet pantsuit with ruffled
white tricot blouse

1968
Red orange minidress with purple and yellow stripes;
yellow plush faux-fur cape and matching hat

1969
Plush faux-fur miniskirt trimmed with orange vinyl; matching mini-length coat with orange vinyl; orange mock turtleneck

1969
Midi dress with taffeta bodice covered in pleated white nylon; black velvet-trimmed empire waist; black taffeta skirt with overskirt of sheer black nylon

the Seventies

1970
Blouse with peasant sleeves and mock turtleneck; multicolored culotte pants; matching minidress

1970
Rose swimsuit with white tie-dye design

1970
Knit pink and yellow dress with empire waist
and pleated skirt; matching jacket

1970
Lamb-look fleecy white coat lined in pink nylon;
pink and black vinyl belt, tote bag, and boots

1970
Pantsuit trimmed in faux fur with gold buttons and clasps;
metallic knit blouse

1970
Metallic knit minidress; mylar metallic blue coat with
blue faux-fur collar and trim; thigh-high boots

1971
One-piece blue tricot swimsuit

1971
Woven spun acrylic plaid jumpsuit; matching plaid jacket
with white faux-fur trim

1971
Pink satin party dress with narrow gold
waistband and handkerchief hem

1971
Raspberry-and-red-dotted swimsuit
and matching wraparound skirt

1971
Velour purple-print peasant dress with laced bodice
and ruffled hem and cuffs

1971
Multicolored mod pants outfit with fringe-trimmed belt and wristbands

1972
Blue denim halter top and cotton
patchwork-print skirt

1973
Gingham dress with sheer sleeves
and black ribbon trim

76

1974
A-line powder blue gown with matching jacket
trimmed in faux fur

1974
Pink-and-white-dotted party dress with ruffled hem
and sheer puffed sleeves

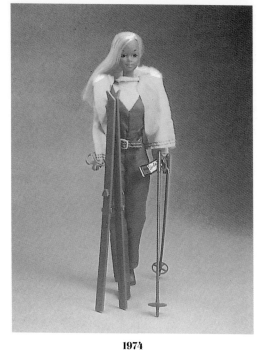

1974
Belted orange snow suit; yellow parka trimmed
with white faux fur

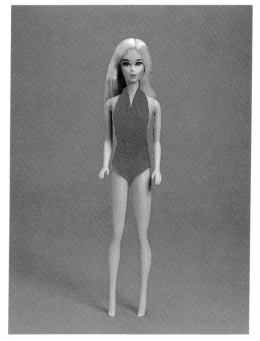

1975
Magenta one-piece halter-style swimsuit

80

1975
Red one-piece swimsuit with square neckline

1975
Red and turquoise tricot halter top and matching
red pants; Indian-print shawl

1976
Navy-and-yellow-striped wraparound tunic top with navy trim and tie belt; solid navy A-line skirt and slacks

1976
Bicentennial dress with fitted blue bodice and red skirt printed with Revolutionary soldiers

1976
White tulle tutu trimmed
in gold braid

1976
Deep rose satin gown with fitted bodice, ecru lace trim, and flounced hem; matching sheer evening shawl

1977
Knit sweater, multicolored vest,
and blue gaucho pants

1977
White tricot blouse with gathered neckline;
straight red skirt with blue topstitching

1977
Brocade one-shouldered evening gown
trimmed with lavender

1977
Empire-style wedding dress with cap sleeves, fitted bodice of white lace
over satin, and white chiffon skirt with front panel of lace and satin

1977
Sleeveless pink satin dress with overdress
of white lace and tulle; lace shawl

1977
Pink jumpsuit with tunic vest

1978
Gold bodysuit, red evening pants, and flowing skirts
in bright red, cherry red, and yellow tulle

1978
Yellow satin gown with gold lamé bodice and sweetheart
neckline; sequin-edged sheer overskirt

1978
Gown with green lamé halter-style bodice
and three-tiered skirt; ruffled boa

1979
Sheer blue voile gown with blue satin evening cape
edged in silver braid

1979
Yellow satin jumpsuit with spaghetti straps;
floor-length hostess skirt

1979
Long sheer floral-print gown with square collar
edged in lace and full sleeves gathered at wrist

1979
Suede-look coat with faux-fur trim; tapestry-print
cotton dress; twisted braid tie belt

1979
Ribbed knit gown with high-necked halter collar trimmed in gold braid;
gold braid belt; faux-fur cape edged in suedelike fabric

1979
Camel double-breasted tricot trench coat
with matching belt

101

1979
Halter-style dress with red bodice
and red floral-print skirt

the
Eighties

1980
Red bodysuit with gold trim and cut-out sleeves;
matching wraparound disco skirt

1980
Gold metallic knit jumpsuit with white taffeta evening coat
trimmed with gold ribbons

1980
White lace blouse lined in white satin gathered at waist;
blue tricot wide-legged pants

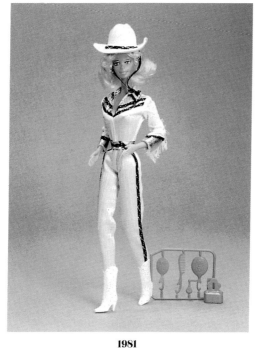

1981
White satin jumpsuit trimmed in silver lamé
and silver-and-black braid

1981
Calico-print Western-style dress with faux-suede-trimmed cuffs, belt, and vest

1981
Sheath-style denim midi skirt; red-and-white calico-print
blouse with Western-style yoke

1982
Gown with ruffled halter-style bodice with high collar and pink ribbon trim; royal blue tricot taffeta skirt with sheer royal blue overskirt

1982
Yellow gown with lace-edged flounced hem,
puffed sleeves, and satin sash

1982
Lace-edged nightgown with narrow ribbon straps;
matching robe with sheer lace sleeves

1982
Embossed silver hostess gown with silver cord trim and tie belt

1984
Blue one-shouldered bodysuit; matching pants; long lace
bouffant skirt trimmed with white faux fur

1984
Strapless fitted lavender tricot bodice and bouffant aqua, lavender, and white printed satin skirt with diagonal ruffle; matching shawl

1984
Pink pinstriped navy jacket with fitted waist and velvet lapels;
hot pink blouse; pale pink velvet jodhpurs

1984
Gray cotton knit ski suit; pink parka with white fun-fur sleeves, hood, and trim

1984
Ruffled pink dress with deep V-neckline,
trimmed in blue

1984
White nylon satin tuxedo; white faux-fur stole;
pink velvet bow tie

1985
Strapless ice blue dress with metallic blue, gold, and magenta highlights on bodice; wraparound skirt trimmed in white faux fur

1985
Ice blue jacket and wraparound midi-length skirt with faux-fur trim; matching faux-fur hood

122

1985
White tricot harem pants; red spandex top;
white pleated wrap

1985
Dress with fitted pink lamé bodice and bouffant skirt of
sheer pink nylon floating over pink sheath underskirt

1986
Strapless red tube-shaped gown with three pleats;
silver mylar sash

1986
Blue-and-white faux-fur-trimmed cape;
blue bodysuit with matching skirt

1986
Electric blue metallic coat, magenta sleeveless tunic,
and lime green lacy spandex leggings

1986
Leopard-print jacket; hot pink spandex top
and stretch pants; yellow sash

1988
Acid-washed waist-length denim jacket; white T-shirt;
gored denim skirt with white topstitching

1988
Waist-length denim jacket with silver studs; pink rib-knit tank top; denim jeans

1988
White cotton dotted top with red rib-knit pants;
red cinch belt

1988
White nubby-knit sweater with royal blue cowl collar and cuffs; royal blue calf-length knit skirt; matching knit cap

1988
Leather-look coat with faux-leopard-skin lapels, cuffs, and trim;
beige blouse with matching scarf; leather-look pants

1988
Oversized white fun-fur coat; gold tissue-lamé
neck bow and turban

1988
Houndstooth-print straight skirt with diamond-plaid blouse;
red brushed tricot jacket with single-button closing

1988
Slim white jacket and skirt trimmed with gold-and-black braid; gold-and-black metallic top with peplum; faux-fur stole and hat

1988
Hip-wrapped dress with gathered skirt;
shiny black faux-fur boa

1988
Deep pink brushed tricot trench coat; pink and white
crepe blouse with matching scarf; green linen pants

1989
White leather-look jacket; black cropped-top with
neon rosettes; four-layered skirt edged in neon

1989
Fuchsia minidress with boat-neck collar; black suedelike
belt with fuchsia satin ribbon

1989
Yellow vinyl miniskirt with diamond-print
sleeveless blouse

1989
One-piece sleeveless sweater dress
with short knife-pleated skirt

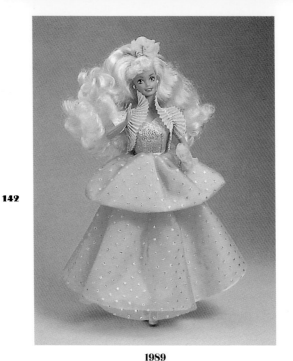

1989
Peach and silver gown with bustier top and peach taffeta skirt
with sheer overskirt; pleated silver metallic jacket

1989
Pink diamond-print dress with dolman sleeves
and metallic skirt; matching hip wrap

the
Nineties

1990
Red sparkle organza gown with layered ruffle skirt, fitted bodice, and ruched sleeves; part of the Happy Holidays series

1991
Bell-skirted black velvet gown; bodice and leg-of-mutton sleeves
trimmed with red and white sequins; part of the Happy Holidays series

1991
Dark pink fringed jacket with blue top and southwestern-style skirt;
cowboy hat and concha belt

1991
Purple denim flared skirt and star-studded vest over
pink knit crop top

1992
Mod-print long-sleeved
micromini-dress

151

1992
Retro-look pantsuit in pink, blue, and purple with matching
purse and stole, accented with pink faux fur

1992
Navy-and-white houndstooth suit with slim-fitting
skirt and yellow silk blouse

1992
Purple and gold lamé asymmetrical gown with
diagonal pleating and one-shoulder swath

1992
Pink glitter-tulle tiered skirt with silver bustier top
and coordinating metallic pink jewelry

1992
White iridescent dress with layered tulle heart-trimmed
skirt and full ruffled collar

1993
Red faux-leather fitted mini-dress accented
with silver belt and earrings

1993
Purple fringed jacket over spangled top and silver lamé fringed skirt; purple cowboy boots

1993
Ruby-red ruffle-tiered gown with fitted bodice and gold accents; part of the Happy Holidays series

1993
Short scarlet dress with pleated ruffle skirt
and appliquéd bolero jacket

1993
White knit suit with gold trim and coordinating
hat, necklace, and purse

1994
Holiday gown with metallic lace skirt and gold lamé overskirt, trimmed in white faux fur; part of the Happy Holidays series

162

1994
Full-skirted raspberry-pink strapless gown of layered sheers
with silver designs and faux-feather hem trim

1994
Pink spangled sheath dress with tightly pleated pink sparkle-sheer cascade

1994
Dressy white faux-leather pantsuit with full peplum; short swing coat with lace sleeves and lapel trim

1995
Black sheath dress and opera cloak accented
with gold lace pattern

1995
Tropical-style bikini, black with floral print
and matching earrings

1995
V-necked bodysuit in hot colors with chartreuse
belt and kneepads

1995
Gown with sheer hombréd overskirt in sherbet colors;
deep pink halter-style bodice with sleeve caps

1995
Faux-fur-trimmed gown with satin bodice and apricot and
lavender skirt accented with silver swirls

1995
Deep pink ruffled short skirt with shiny pink
waistcoat-style bodice

1995
White cotton blouse under school-motif-printed
black cotton dress

1995
Holiday gown with holly-print overskirt, lace underskirt, poufed sleeves, and holly-trimmed collar; part of the Happy Holidays series

1996
Victorian-style velvet sleeved cloak over tiered gold damask dress; white faux-fur collar, cuffs, and hat; part of the Happy Holidays series

1996
High-waisted mini-dress with pink
velour top and yellow skirt

1996
Sleeveless white silk gown with full-gore bias-cut skirt, accented with silver

1996
Bell-skirted white and gold gown with
fitted white bodice and flared peplum

177

1996
Black sheath dress with paisley design and
poufed red satin sleeves

1996
Suit in lavender wool with gold buttons
and chain belt

179

1997
Lavender satin suit dress with white
faux-fur collar and cuffs

1997
Yellow rain slicker with coordinating hat;
plaid lining to match umbrella

1997
Black jacket over striped shirt and
slim-cut white trousers

1998
Navy bouclé coat with tan faux-fur
cuffs and shawl collar

1998
Black-and-white pinstriped suit
with slim-cut trousers

1998
Peach satin fitted suit with long-line jacket and slim skirt;
leopard-print collar and hat

1998
Sage-green floor-length slit skirt; pink satin strapless bodice; pink moiré jacket with deep shawl collar

1998

Left to right: Orange flared-sleeved jacket with black pants;
pink spangled mini-dress with sheer sleeves;
chartreuse cropped T and black leggings with chartreuse side stripe

1998
Blue corduroy bib overalls with slim-striped mock turtleneck

Appendices

Nationalities Barbie Doll Has Represented

(International and Dolls of the World Collections)

1980	Italian Barbie, Parisian Barbie, Royal Barbie (of England)
1981	Scottish Barbie, Oriental Barbie
1982	Eskimo Barbie, East India Barbie
1983	Spanish Barbie, Swedish Barbie
1984	Irish Barbie, Swiss Barbie
1985	Japanese Barbie
1986	Greek Barbie, Peruvian Barbie
1987	German Barbie, Icelandic Barbie
1988	Korean Barbie, Canadian Barbie
1989	Russian Barbie, Mexican Barbie
1990	Nigerian Barbie, Brazilian Barbie
1991	Malaysian Barbie, Czechoslovakian Barbie
1992	Jamaican Barbie
1993	Native American Barbie, Australian Barbie
1994	Dutch Barbie, Chinese Barbie, Kenyan Barbie
1995	Polynesian Barbie, German Barbie, Irish Barbie
1996	Japanese Barbie, Indian Barbie, Mexican Barbie, Norwegian Barbie, Ghanaian Barbie
1997	French Barbie, Russian Barbie, Arctic Barbie, Puerto Rican Barbie
1998	Chilean Barbie, Polish Barbie, Thai Barbie, Native American Barbie

Barbie Doll's Careers

1959	Teenage Fashion Model
1961	Ballerina, Registered Nurse, American Airlines Stewardess
1963	Graduate, Career Girl
1964	Candy Striper Volunteer
1965	Astronaut, Fashion Editor, Student Teacher
1966	Pan Am Stewardess
1973	Surgeon
1975	Olympic Athlete: Downhill Skier, Figure Skater, Gymnast
1984	Aerobics Instructor (Great Shape Barbie)
1985	Business Executive (Day-to-Night Barbie), Dress Designer, TV News Reporter, Veterinarian, Teacher
1986	Astronaut, Rock Star
1988	Doctor
1989	UNICEF Ambassador, Doctor, Army Officer, Dancer on a TV Dance Club Show
1990	U.S. Air Force Pilot, Rock Star, Summit Diplomat, Ice Capades Star
1991	Music Video Star, Naval Petty Officer
1992	Marine Corps Sergeant, Rap Musician, Rollerblade In-Line Skater, Teacher, Chef, Businesswoman, Doctor, Presidential Candidate
1993	Police Officer, Army Medic, Radio City Music Hall Rockette, Baseball Player
1994	Pediatrician, Astronaut, Scuba Diver, Air Force Thunderbird Squadron Leader, Artist
1995	Teacher, Lifeguard, Firefighter
1996	Veterinarian, Engineer, Olympic Gymnast
1997	Dentist, Paleontologist, Boutique Owner
1998	Musician, Olympic Skater, Pilot, Professional Race Car Driver

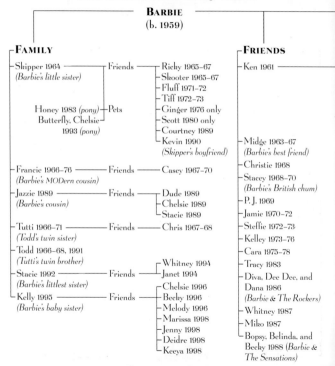

- Family — Little Brother Tommy 1997
  - Friends
    - Allan 1964–65, 1991 *(Midge's boyfriend)*
    - Brad 1970–72 *(Christie's boyfriend)*
    - Curtis 1975 only *(Cara's boyfriend)*
    - Todd 1983 *(Tracy's fiancé)*
    - Derek 1986
    - Steven 1988 *(Christie's boyfriend)*

- Teresa 1988
- Midge 1988
- Kayla and Devon 1989 *(Dance Club)*
- Kira 1990 *(Wet 'N' Wild)*
- Nia 1990 *(Western Fun)*
- M. C. Hammer 1991 *(Celebrity friend)*
- Tara Lynn 1993 *(Western Stampin')*
- Kayla, Shani 1994 *(Locket Surprise)*
- Becky 1997

- Prancing Horse 1994 *(horse)*
- Tropical Sea Horse 1995 *(horse)*
- Colley 1996 *(dog)*
- Calico 1996 *(cat)*
- Nibbles 1996 *(horse)*
- Ginger 1997 *(dog)*

Pets

- Dancer 1971–72 *(horse)*
- Beauty 1980–83 *(Afghan)*
- Beauty and Pups 1982–83
- Dallas 1981 *(horse)* ⎫
- Midnight 1982 *(horse)* ⎬ Family
- Dixie 1984 *(baby palomino)* ⎭
- Prancer 1984 *(Arabian stallion)*
- Fluff 1983 *(kitten)*
- Prince 1985 *(Poodle)*
- Blinking Beauty 1988 *(White horse)*
- Sun Runner 1990 *(horse)*
- All American 1991 *(horse)*
- Sacha 1992 *(puppy)*
- Honey 1992 *(kitten)*
- Rosebud 1992 *(horse)*
- Tag Along Wags 1993 *(Puppy)*
- Tag Along Tiffy 1993 *(Kitty)*
- Mitzi Meow 1994 *(kitty)*

MINIS**ERIES** TITLES AVAILABLE FROM ABBEVILLE PRESS

- **CATS UP CLOSE**
  0-7892-0510-6
  $5.95

- **ELVIS: HIS LIFE IN PICTURES**
  0-7892-0509-2
  $5.95

- **HORSES**
  0-7892-0526-2
  $5.95

- **TROPICAL COCKTAILS**
  0-7892-0554-8
  $5.95

- **WEDDINGS**
  0-7892-0524-6
  $5.95